Think-A-Lot-Tots:
The Neuron

By Dr. Thomai Dion

ISBN-13: 978-1534968660

To my little guy.
My son, and my sun.

Our bodies can do a lot of amazing things.

Some can run very fast.
Or sing very beautifully.
Or draw very skillfully.

Our bodies can move
in many different ways.
What types of things can you do?

How do we smile?
Or wave hello?
Or stomp in puddles?
How does our body know how to move?

Inside of our head is the brain.
The brain is very important.

It tells our body how to move!

The brain moves our body by sending messages.
Like sending a letter in the mail.
Or calling a friend on the phone.

A message may tell our arm to wave.
Or our feet to stomp.
Or our face to smile.

The brain has helpers that carry messages to our arms, feet, face, and other places. These helpers are called neurons. What is a neuron?

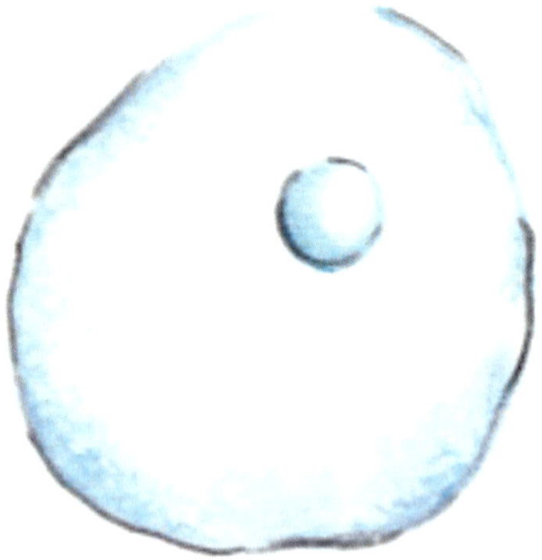

A neuron is a type of cell in our body that sends messages.
This part of the neuron is very big.
This is where the nucleus is.

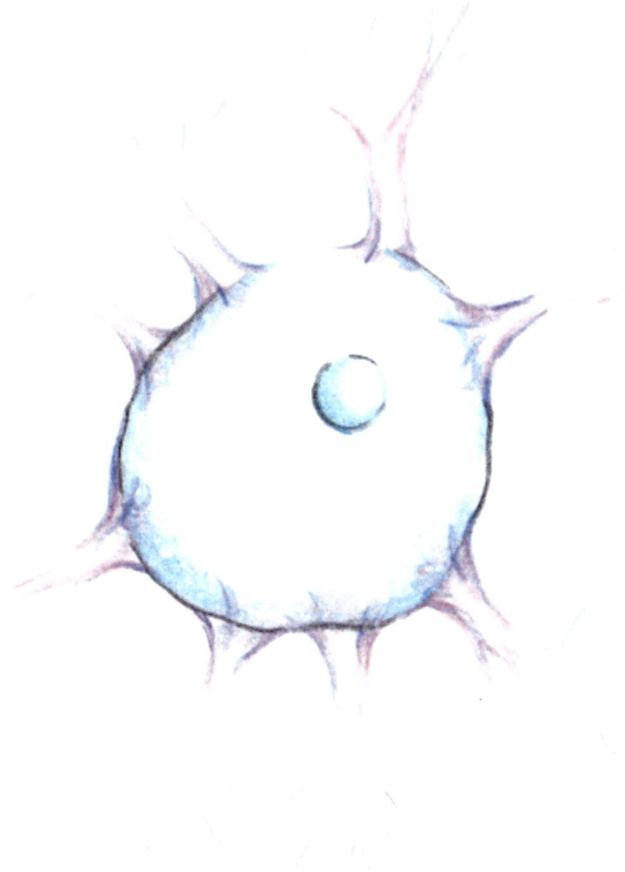

These are dendrites.
They are around the nucleus.
They look like big tree branches!

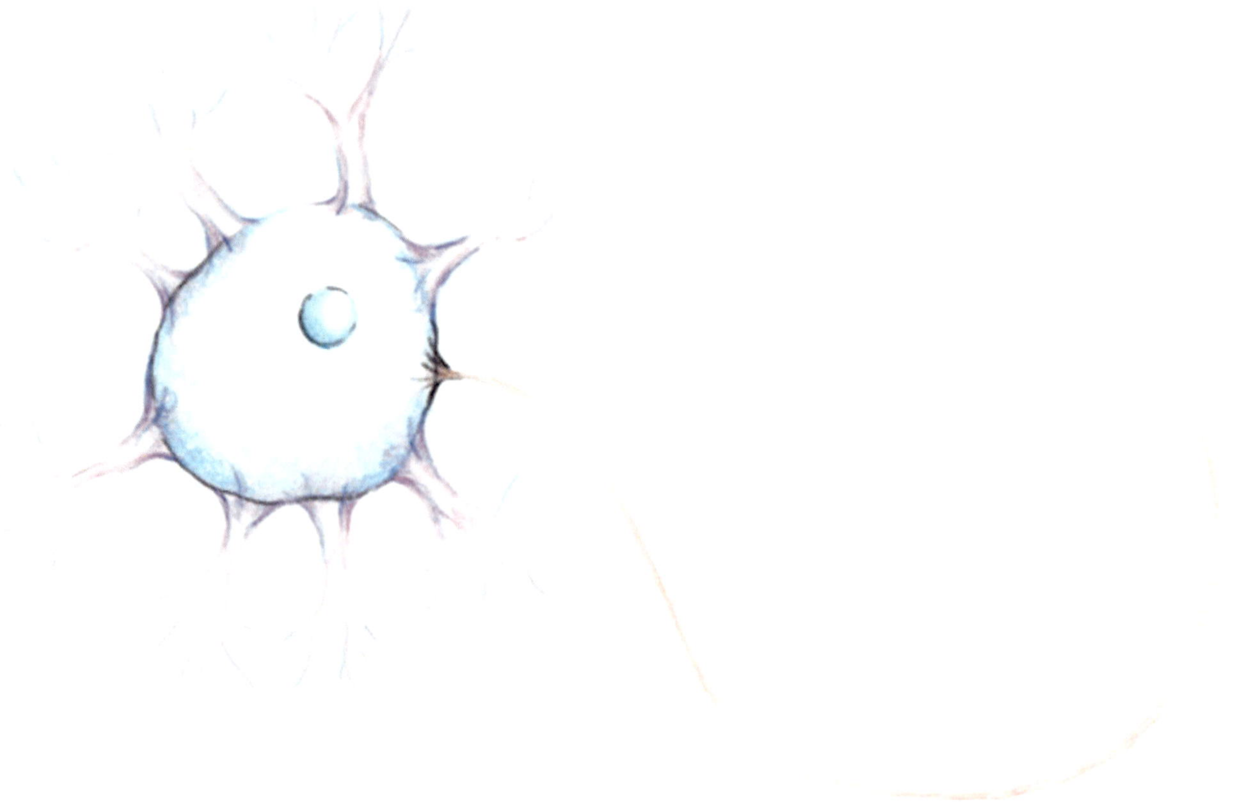

This is the axon.
It is long and looks like a big tail.

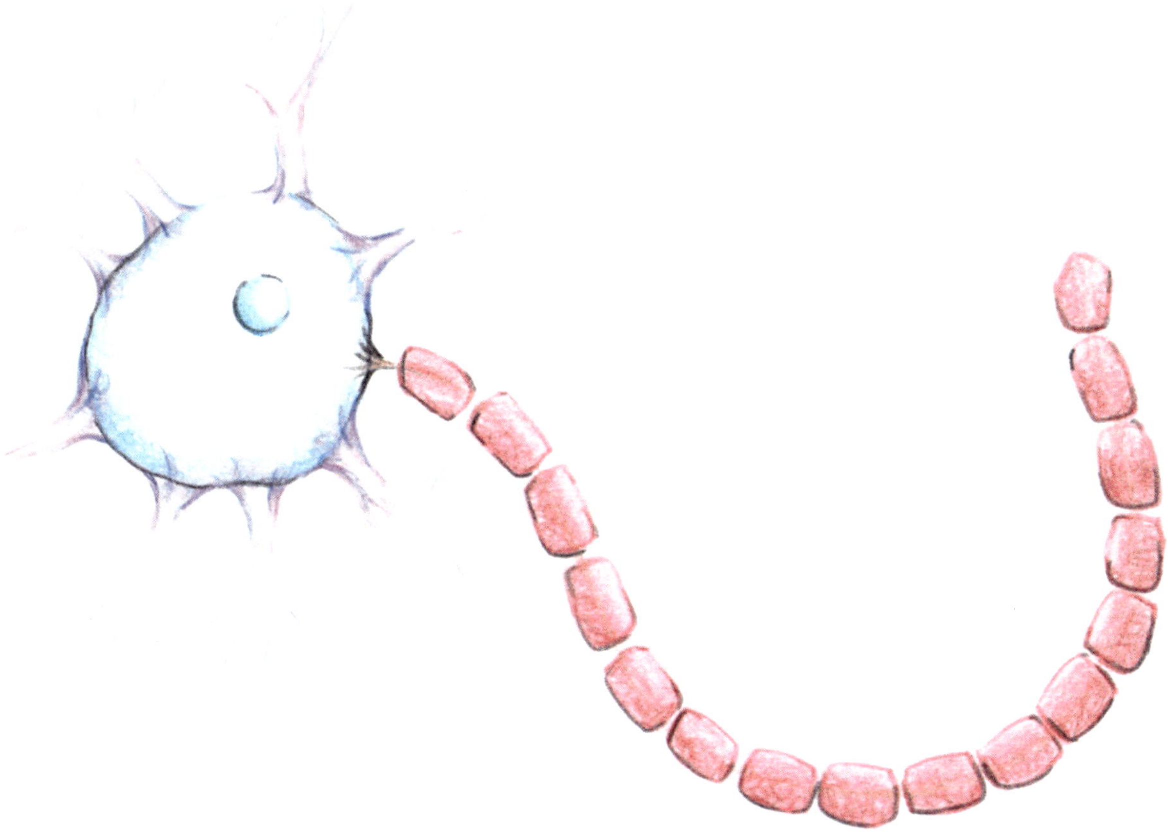

The axon is covered by the myelin sheath. Now it looks like a big necklace!

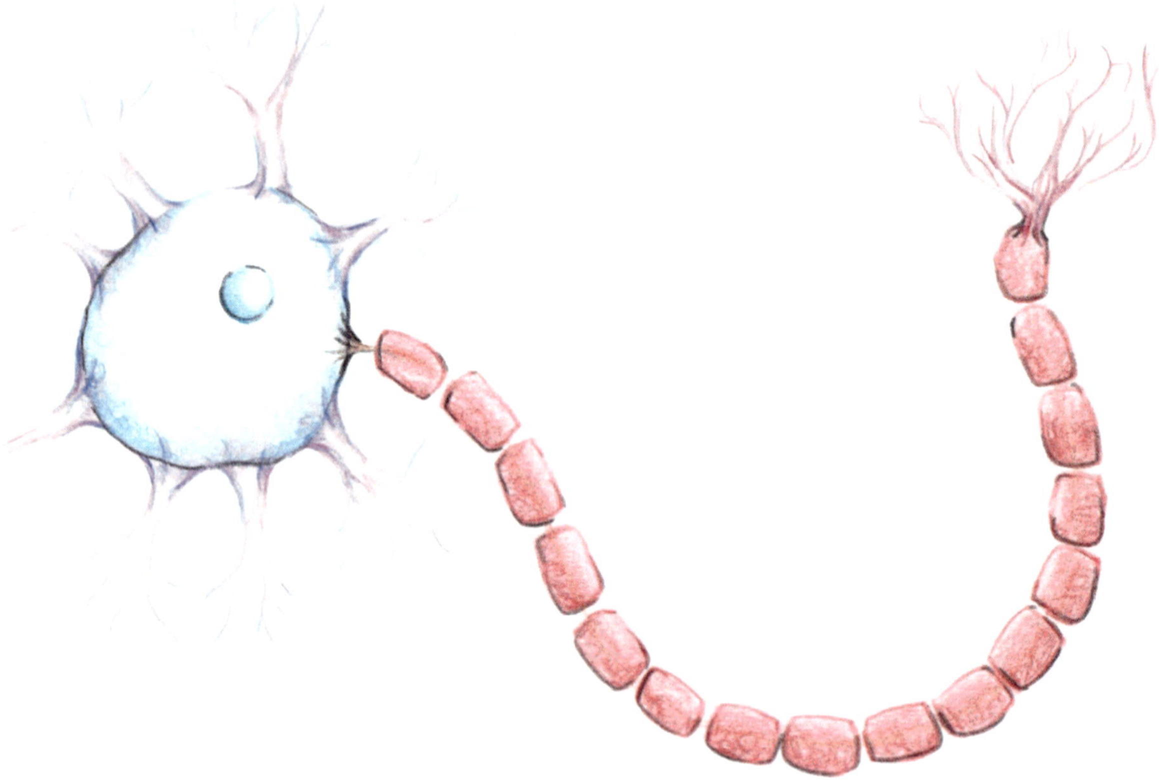

At the end of the axon are the terminals.
They look very bushy.
Just like another tree!

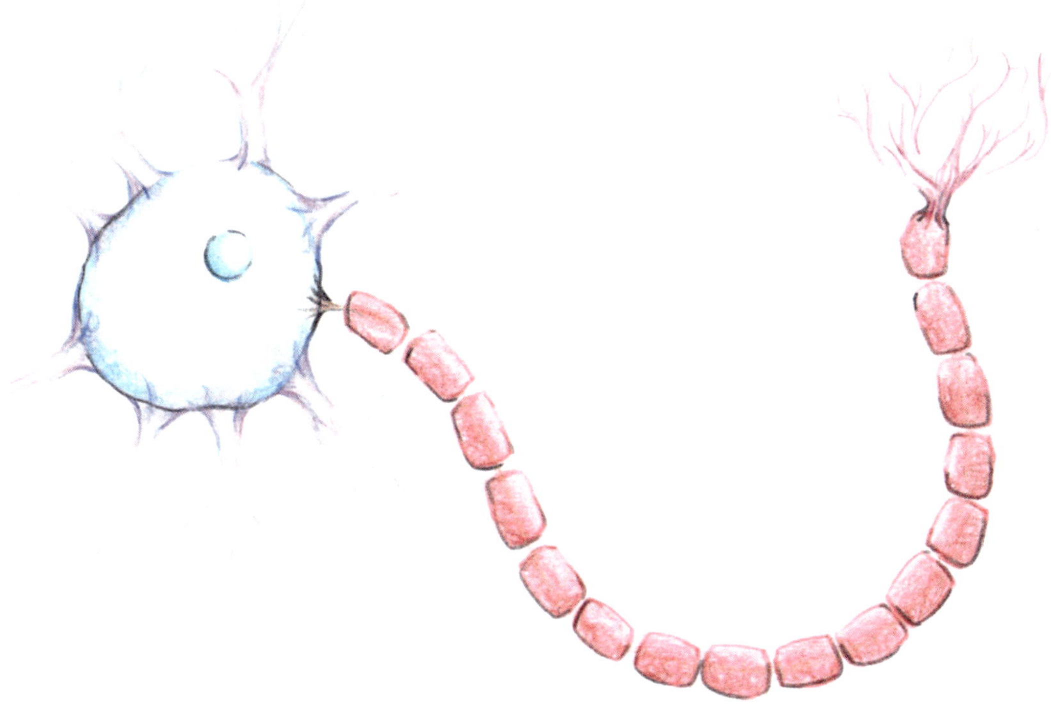

Now you know the parts of a neuron!

How does the neuron
carry messages for the brain?

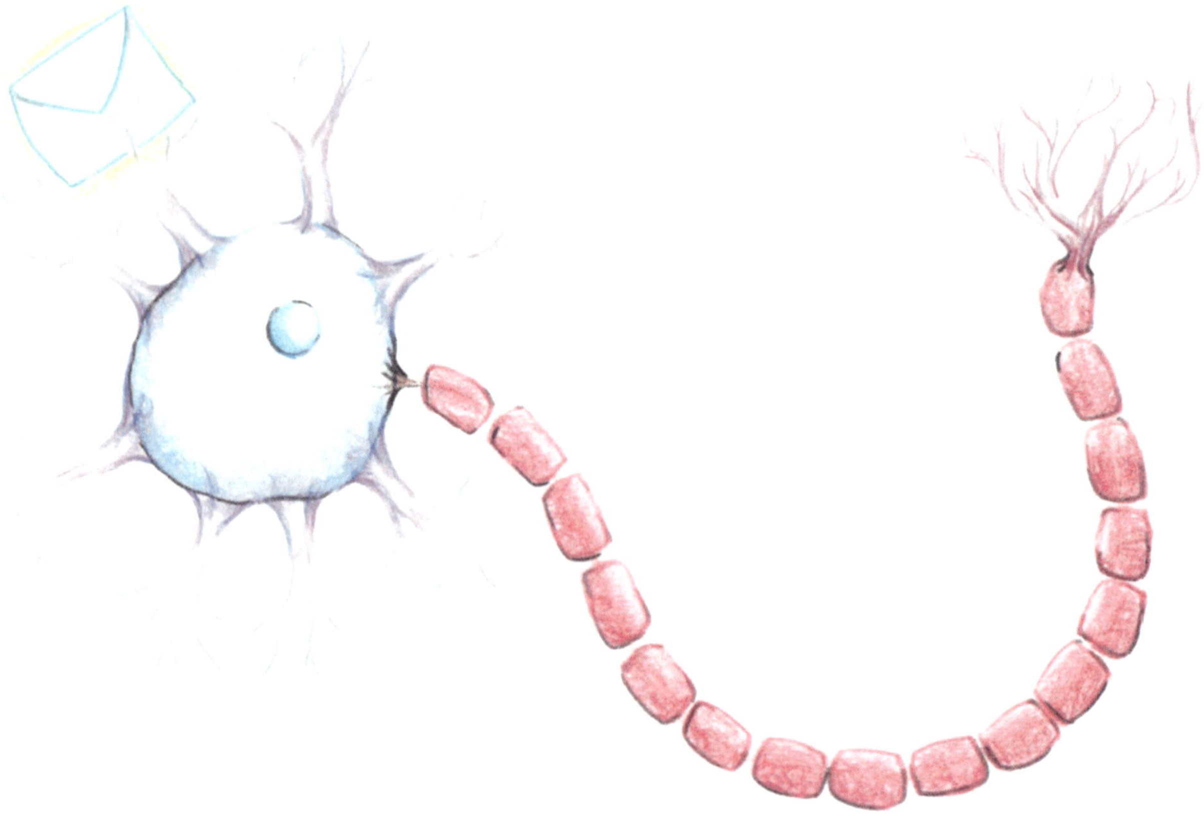

Here is a message, dendrites!
It is from the brain.
Thank you for taking this.
Make sure it goes where it needs to!

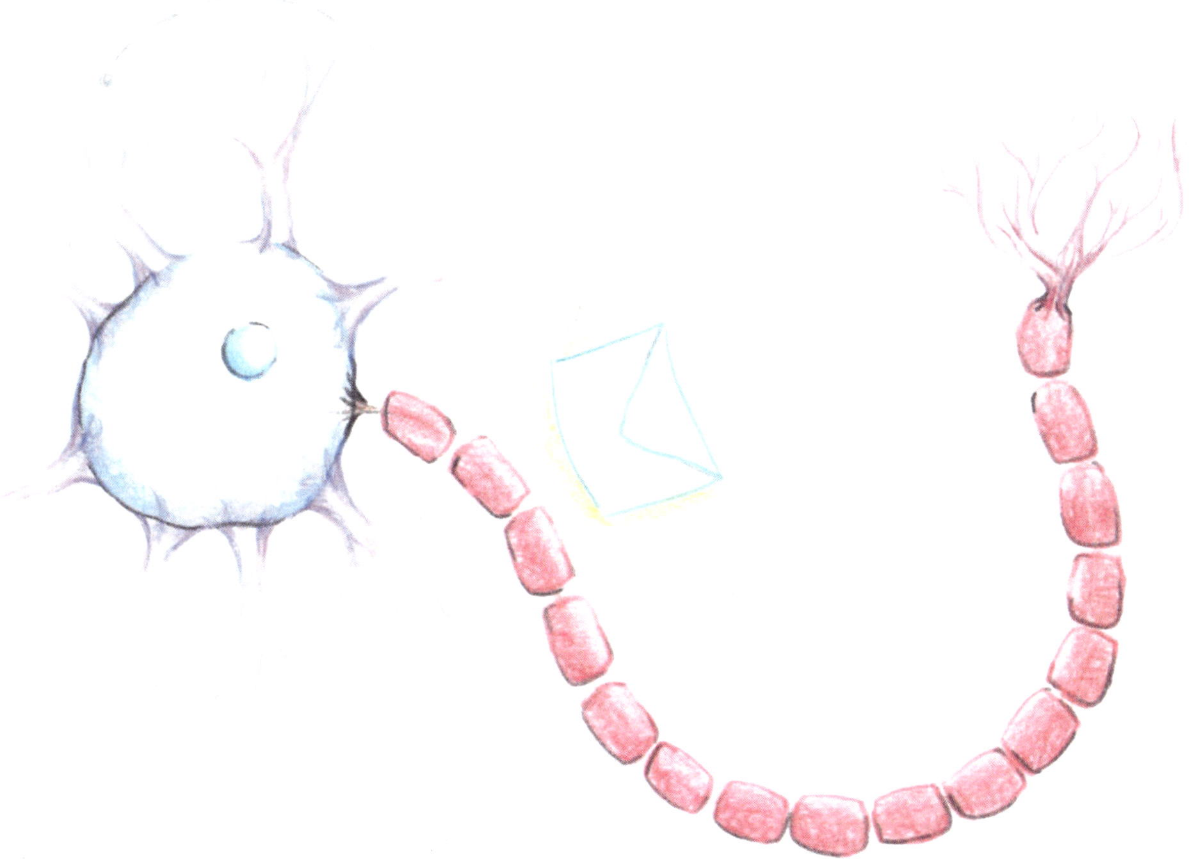

It is now your turn to take the message from the dendrites.
Thank you for your help, axon!

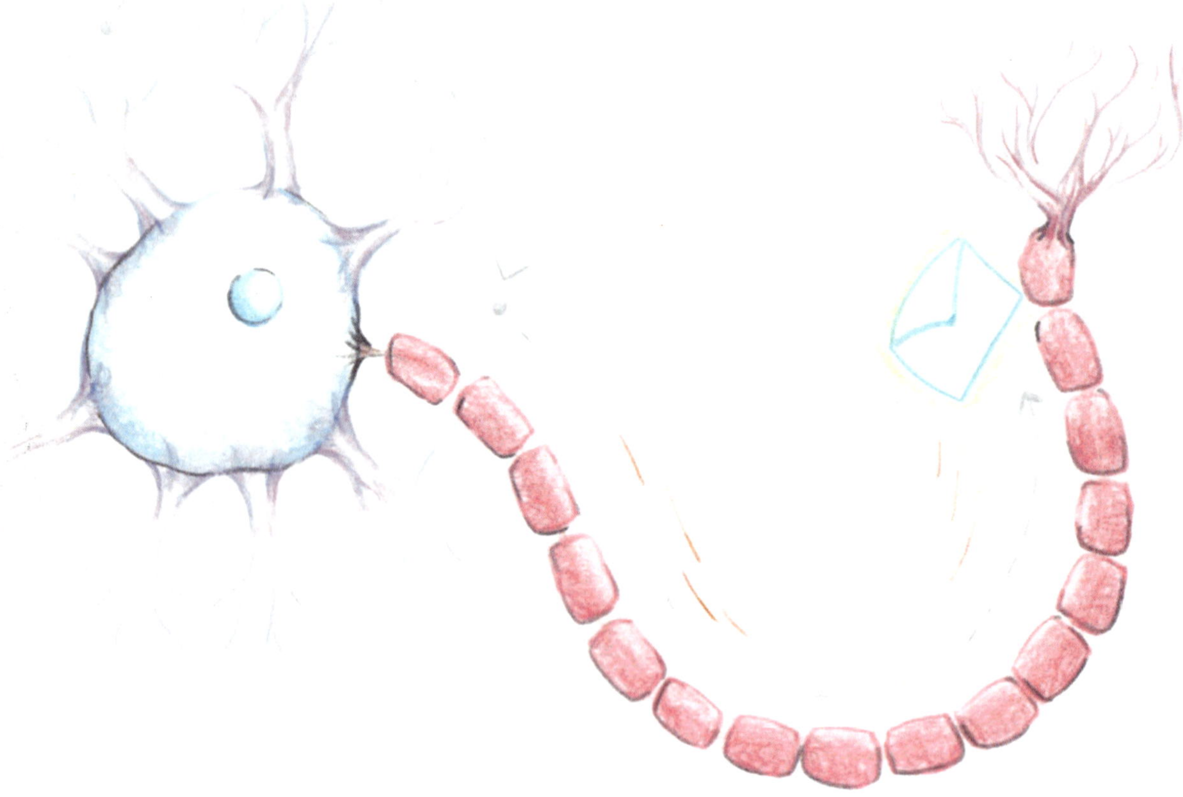

The myelin sheath lets the axon carry the message very quickly.

Zoom!

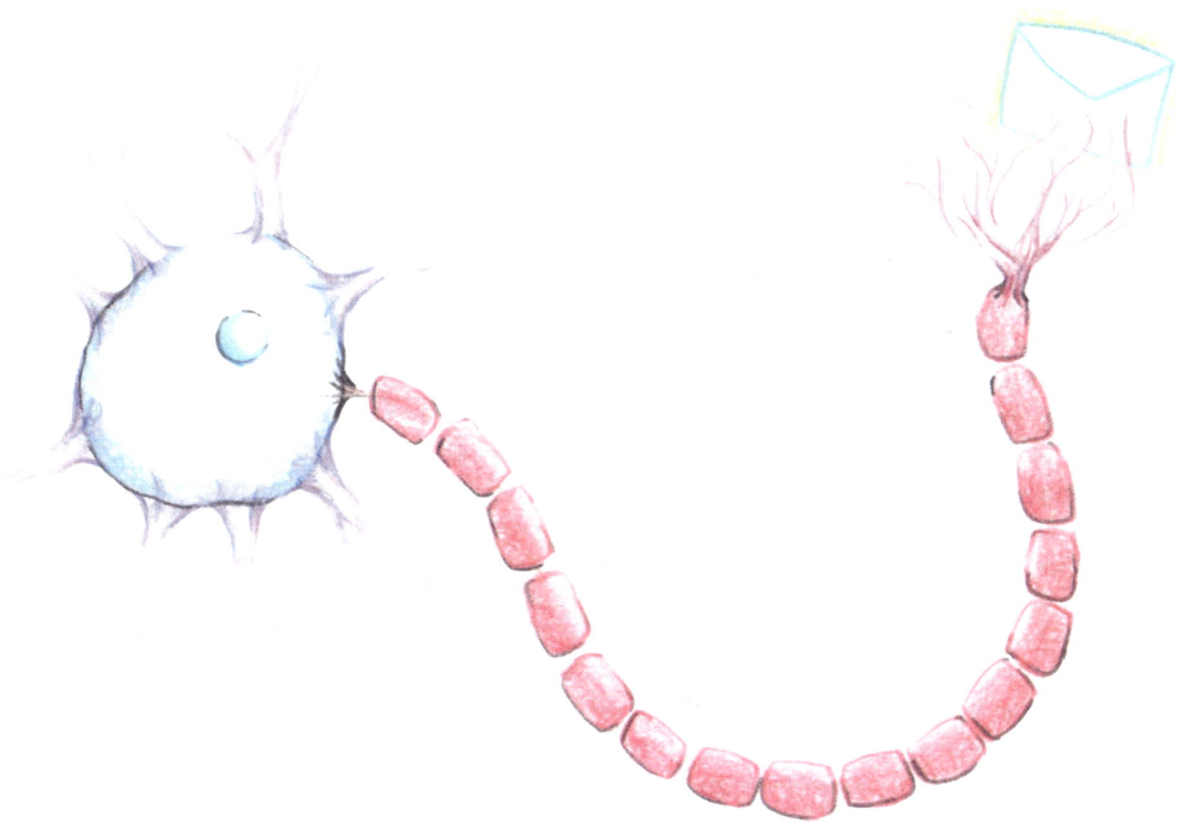

We have reached the end of the axon!
Here is the message, terminals.
What do they do with the message?

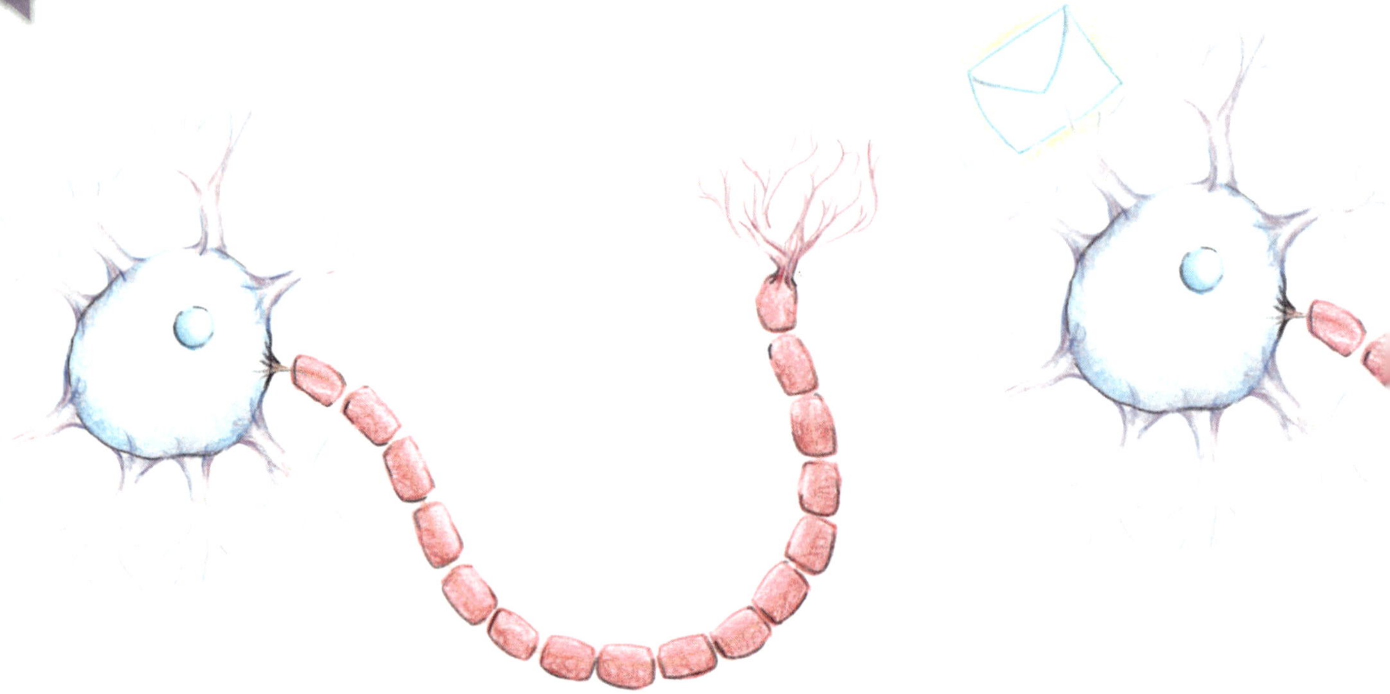

The terminals find another neuron to help carry the message!
The other neuron's dendrites will take the message from the axon's terminals.

All of the neurons help each other
carry the brain's message
to where it needs to go in the body.
Great job, everyone!

ABOUT THE AUTHOR

Thomai Dion is a pharmacist and mother to two very inquisitive, energetic, and hands-on analytical thinkers. She obtained her doctorate from the University of Rhode Island and believes it is never too early to start learning.

For more books that promote children's S.T.E.M. education, visit Dr. Dion's website:

www.tdthesciencemom.com

86355398R00018

Made in the USA
Lexington, KY
11 April 2018